Foundations of Matter

by Christine Caputo

Table of Contents

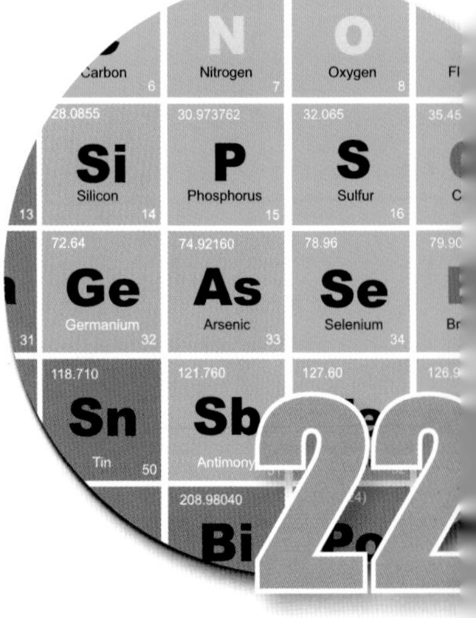

4
20
6
WHAT ARE ELEMENTS MADE OF AND HOW ARE THEY ARRANGED IN THE PERIODIC TABLE?

THE Great Wall

If you travel to China, you will see one of the most amazing structures ever built—the Great Wall of China. Many consider it one of the greatest wonders of the world. When first constructed, this great wall was made up of separate sections. Each section protected a different Chinese state. Guards on watchtowers used fire or smoke signals to warn of approaching danger.

In 221 B.C., the Qin dynasty united the Chinese states. The leaders ordered the walls to be connected. After many years in which millions of Chinese workers labored, the completed wall stretched about 6,700 kilometers (4,163 miles) from east to west across China. At an average height of almost 8 meters (26.2 feet), the wall snakes through grasslands, over mountains, and across deserts.

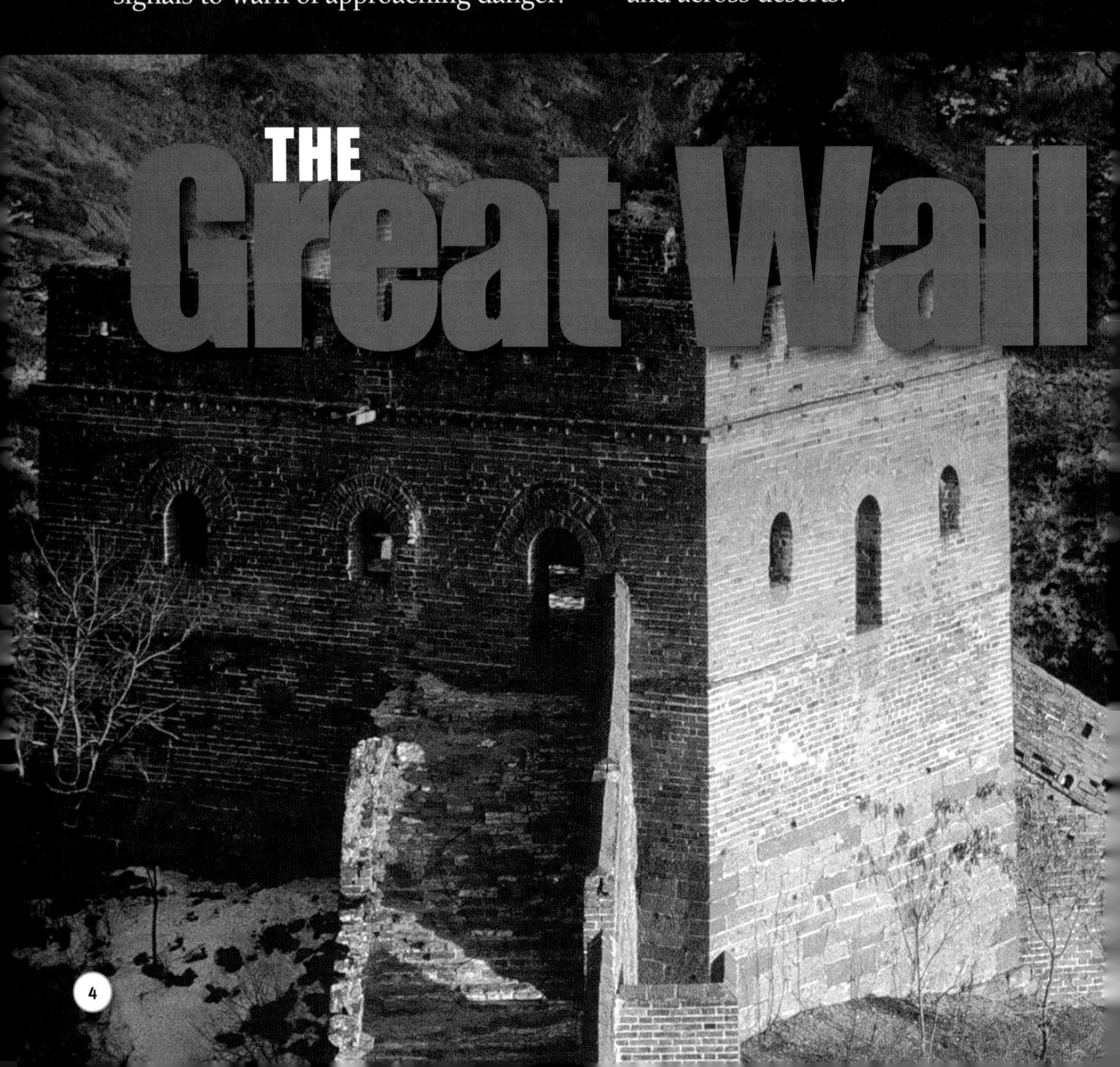

If you look at the Great Wall from a distance, it appears to be one solid structure. Look more closely, however, and you will see that it is made up of many smaller parts. Early sections of the wall were made from cut stones and mud shaped like boxes. You can also see heavy bricks made from clay and granite. It is estimated that the number of bricks and stones used to build the Great Wall could build a wall 5 meters (16.4 feet) tall and 1 meter (3.3 feet) wide that would stretch around the entire planet!

In many ways, the Great Wall of China is like the matter that fills the world. You, your home, and the planet on which you live are all examples of matter. What does matter have in common with the Great Wall of China?

# of China

THE GREAT WALL OF CHINA HAS BEEN SAID TO WIND ACROSS THE COUNTRY LIKE A GIANT DRAGON.

CHAPTER 1

Development of the Atomic Theory

How did the modern theory of the atom develop over time?

Perhaps you have heard the saying, "There is more than meets the eye." In other words, there is more to something than you can see at first glance. This is especially true when it comes to **matter**. Matter is anything that has mass and takes up space (has volume). Your books, your desk, and your pencil are all matter.

Matter is made up of building blocks called **atoms**. An atom is incredibly small. A single copper penny may contain about 28 sextillion (28,000,000,000,000,000,000,000) copper atoms! Although atoms are very small, scientists today know a great deal about them. How is this possible? The story begins more than 2,000 years ago.

The Root of the Meaning

The word *atom* comes from the Greek word *atomos*, meaning "indivisible."

In ancient Greece, philosophers debated about the makeup of all the matter in the world around them. The Greek philosopher Democritus (dih-MAH-krih-tus) had a theory. He believed you could cut matter into smaller and smaller pieces and, at some point, reach a particle that could no longer be cut. He named this particle an atom. Democritus suggested that atoms were small, solid particles made up of the same material but shaped in different ways. He also proposed that atoms could combine in different ways.

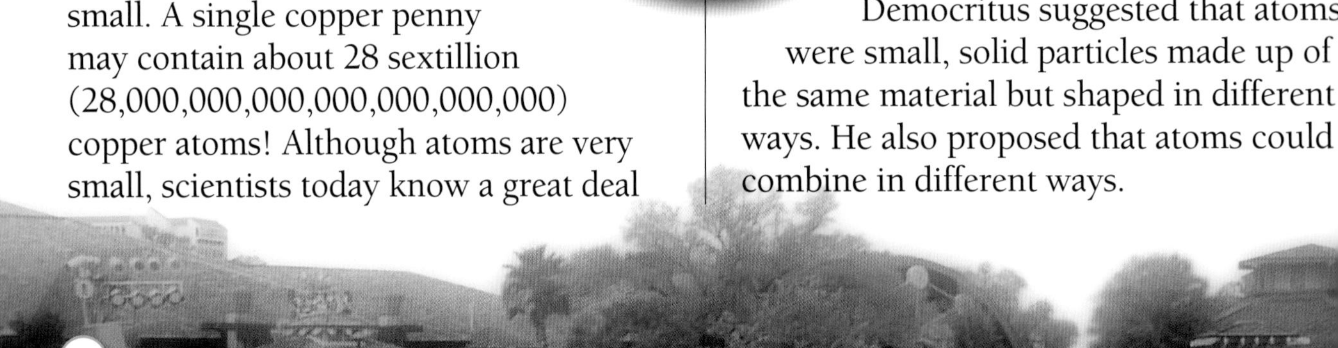

Essential Vocabulary

atoms	page 6
electron clouds	page 12
electrons	page 10
matter	page 6
neutron	page 11
nucleus	page 11
protons	page 11

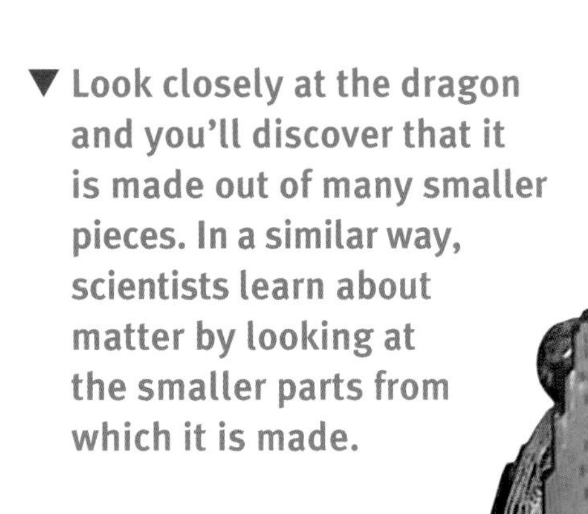

▼ Look closely at the dragon and you'll discover that it is made out of many smaller pieces. In a similar way, scientists learn about matter by looking at the smaller parts from which it is made.

ARISTOTLE

An Alternative Theory of Matter

Another Greek philosopher, Aristotle (A-rih-stahl-tul), disagreed with Democritus. Aristotle believed that the universe was made up of four different substances that he called elements—earth, fire, air, and water. He proposed that these elements mix together to form every known substance. Because Aristotle was a popular thinker of the time and because atoms are invisible, people tended to believe Aristotle rather than Democritus.

DALTON

An Atomic Theory of Matter

Aristotle's ideas were not challenged for more than 2,000 years. It was not until the nineteenth century that a new theory emerged. The British scientist John Dalton reasoned that substances were made up of individual particles. These particles were similar to the idea of the atom that had been proposed by Democritus. Dalton conducted many experiments to investigate his idea. Based on those experiments and other observations, Dalton proposed the following atomic theory in 1803:

Elements are made up of tiny particles called atoms.

- All matter is made up of particles that cannot be divided, created, or destroyed.
- Atoms of the same element are identical: they have the same mass and properties. Atoms of different elements have different masses and properties.
- Two or more kinds of atoms can join together to form chemical compounds.

Dalton's atomic theory is the foundation for the modern atomic theory that scientists accept today. In the years after Dalton's work, other scientists developed and refined his theory.

◀ **In about 350 B.C., Aristotle proposed the idea that everything on Earth could be described in terms of four elements: earth, fire, water, and air.**

▲ John Dalton used these wooden balls to represent his idea of atoms. Each ball was a solid object that could not be divided into parts.

Science and Math

Scientific Notation

When describing very small or very large measurements, scientists use a shorthand method known as scientific notation. In scientific notation, a number is written as a coefficient multiplied by a power of 10. Remember the number of copper atoms in a penny that you read about earlier?

$$28{,}000{,}000{,}000{,}000{,}000{,}000{,}000 = 2.8 \times 10^{22}$$

THOMSON

Looking Inside the Atom

The next breakthrough in understanding the atom came in 1897. The British scientist J.J. Thomson wanted to find out more about the nature of matter and atoms. To do so, he conducted experiments with electricity. Electricity is made up of moving electric charges. Electric charges can be positive or negative. Opposite electric charges attract each other, and like electric charges repel each other.

Thomson worked with a glass tube from which most of the air had been removed. When electricity flowed in the circuit, a beam passed through the tube. Thomson observed that the beam was pulled toward a positively charged plate attached to the tube. Because opposite charges attract, he concluded that the beam must be made up of particles with a negative charge. Thomson reasoned that the particles were parts of atoms. These particles became known as **electrons**. Electrons (ih-LEK-trahns) are negatively charged particles within atoms. Thomson's bold suggestion was the first time scientists had considered the existence of anything inside an atom.

THE ROOT OF THE MEANING

The word *electron* comes from the Greek word *ēlektron,* meaning "amber." Amber is a yellow resin in evergreen trees that becomes fossilized. The Greeks knew that when amber was rubbed with a dry cloth, it would produce static electricity.

◀ **Using a cathode ray tube, Thomson discovered electrons. Because he compared the arrangement of electrons in an atom to raisins in plum pudding, a popular food in his time, his theory came to be known as the plum-pudding model of the atom.**

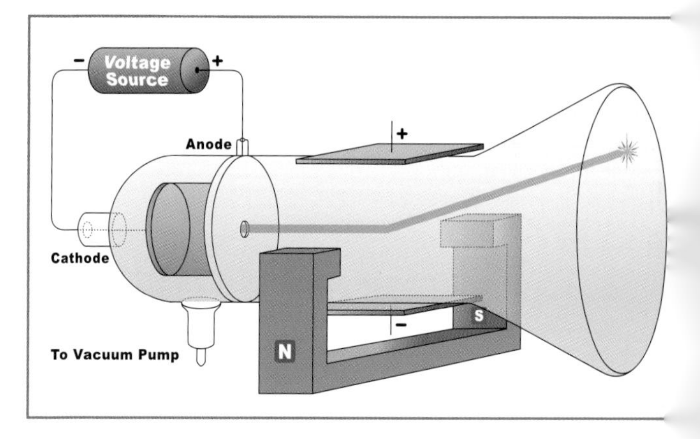

RUTHERFORD

A Model of the Atom

J.J. Thomson had discovered the part of the atom called the electron. In 1909, a student of Thomson's named Ernest Rutherford identified another part of the atom. Rutherford conducted an experiment involving alpha rays, which are made up of positively charged particles. Rutherford aimed a beam of alpha particles at a very thin sheet of gold foil. He surrounded the foil with a material that would glow when struck by an alpha particle.

Based on Thomson's model of electrons spread evenly throughout a positive sphere, Rutherford expected most of the alpha particles to pass right through the foil—which they did. What he did not expect was that some particles would be bent to the left and right, and some would bounce right back to where they started. Because like charges repel, Rutherford deduced that the positive charge of an atom must be concentrated in the **nucleus** (NOO-klee-us), or center. The particles with positive charge later became known as **protons** (PROH-tahns).

From his work, Rutherford developed a model of the atom. At the center of the atom was the nucleus. Orbiting, or circling, the nucleus were the negatively charged electrons. In 1932, the scientist James Chadwick discovered a third atomic particle. A **neutron** (NOO-trahn) has about the same mass as a proton, but no electric charge. It is said to be electrically neutral. Neutrons are located in the nucleus along with protons.

Everyday Science

Static Electricity

Have you ever gotten a shock when touching a doorknob or had your hair stand on end after pulling off your hat? If so, you have experienced static electricity. Electric charges can be moved from one object to another by rubbing. Static electricity is an imbalance (unequal distribution of electric charges). The object being rubbed may have lost electrons to the object that was rubbing it. As the electrons move back to where they came from, an electric discharge occurs. Electric discharges can range from tiny shocks to bolts of lightning.

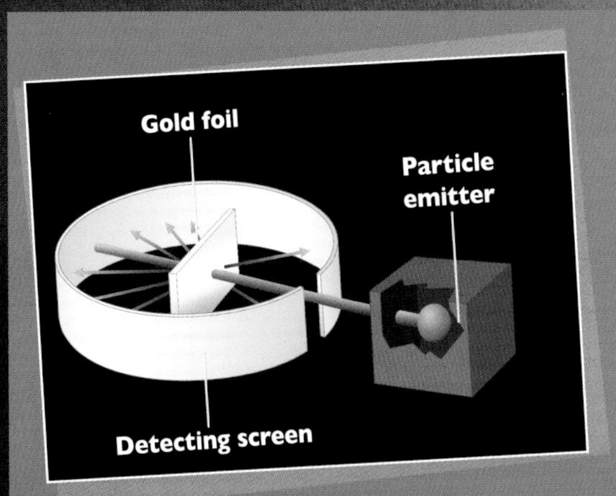

▲ This diagram shows how the alpha rays struck the gold atoms of the foil. Because the nucleus is so small, most of the alpha rays passed right through the empty space of the atom. Those that passed near the nucleus were bent off course. Those that hit the nucleus directly were bent in the opposite direction.

BOHR

Energy Levels

In 1913, the Danish scientist Niels Bohr set out to improve on Rutherford's model of the atom. Bohr developed a new model in which electrons orbit the positively charged nucleus in levels. The level in which an electron is found depends on the amount of energy it has. According to Bohr, electrons could absorb energy and move to a higher energy level. When the electron returns to its original lower level, it gives off energy. Electrons would not be found between energy levels.

ELECTRON CLOUD THEORY

The Modern Theory

In the mid-1920s, scientists determined that the atom was more complex than Bohr's model indicated. It turns out that electrons sometimes behave like waves instead of particles. This made it impossible to determine the exact location of an electron at any point. All that could be calculated was the likelihood that an electron was in a certain location. The regions in which electrons are likely to be found are known as **electron clouds**.

checkpoint

Read More About It

Scientific discoveries sometimes make it possible to develop new technology. Other times, new technology makes it possible to make scientific discoveries. Conduct additional research to find out how technology was related to the development of the atomic theory. Present your findings to the class.

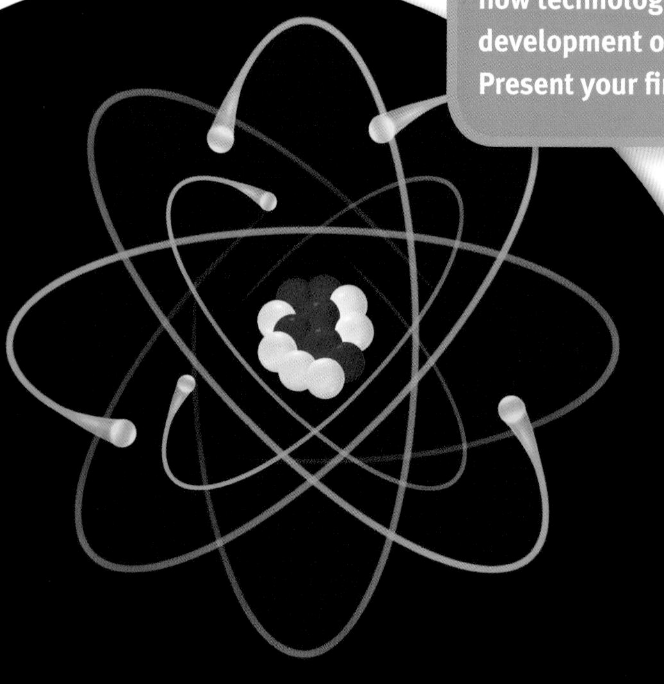

▲ The diagram shows the Bohr model for carbon. It has six protons and six neutrons in the nucleus. You can see six electrons traveling in energy levels around the nucleus

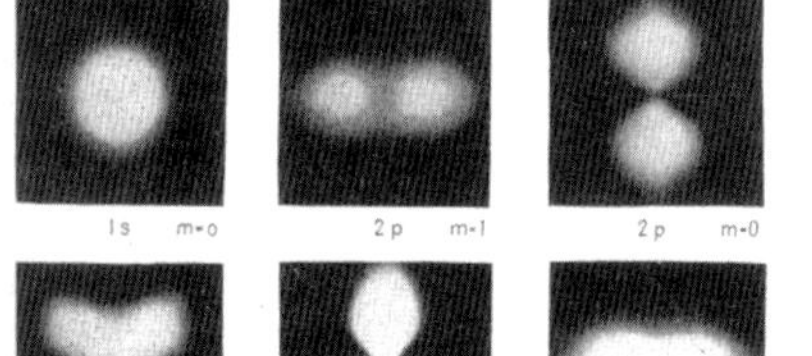
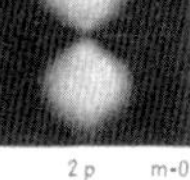
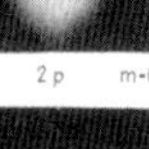
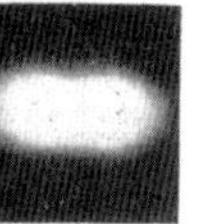
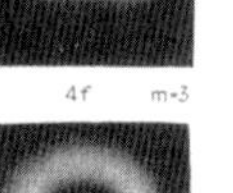
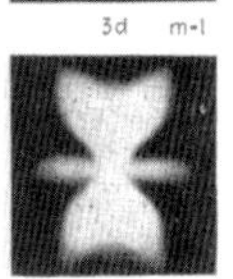
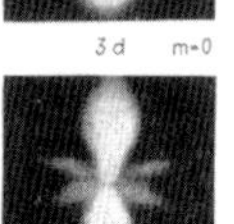
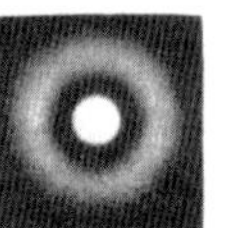
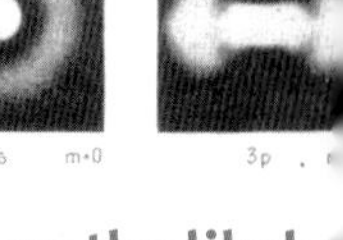

▲ An electron cloud shows the likely locations of electrons as they move around the nucleus of an atom.

Summing Up

- The theory of the atom has been developing and changing for more than 2,000 years. Democritus first proposed the idea of the atom, but the idea was not further developed until Dalton's work in the nineteenth century.
- Thomson then discovered the electron, which showed for the first time that atoms are made up of smaller particles.
- Rutherford went on to show that the positive charge and mass of an atom are concentrated in the nucleus.
- Bohr proposed that electrons orbit the nucleus in energy levels.
- The electron cloud theory (the modern theory) showed that the motion of electrons can better be described by clouds.

Putting it All Together

Choose from the following research activities. Work independently, in pairs, or in a small group. Share your responses with the class. Listen to other groups present their answers.

1. The atomic theory has changed over time. Work with a partner to make a time line that describes important discoveries and updates to the theory. Include dates and pictures.

2. Choose one of the scientists described in this chapter. Create an oral presentation that describes the scientist's contributions to the development of the atomic theory. Conduct additional research to learn more about the scientist. Include diagrams as appropriate in your presentation.

3. Theories are important foundations of science. Write a paragraph explaining what a theory is and why theories might change over time. Tell whether you think there is still value in a theory that turns out to be incorrect.

Chapter 2

Describing Atoms

How are atoms described in terms of their protons, neutrons, and electrons?

As you read in Chapter 1, the current understanding of the atom developed over time through the work of many scientists. The theory may continue to change in the future as new information is learned. Currently, however, the theory of the atom suggests that every atom is made up of smaller particles. These particles, known as subatomic particles, include the proton, neutron, and electron. The table compares the particles in terms of charge, mass, and location.

Scientists describe atoms in a number of ways, including atomic number, mass number, and **atomic mass**. All of these numbers provide information about the number of protons, neutrons, and electrons in an atom.

Atomic Number

The number of protons in the nucleus of an atom is the **atomic number**. All atoms with the same atomic number belong to the same **element**. Oxygen, carbon, and copper are examples of elements. The atomic number of oxygen is 8. That means that every oxygen atom has eight protons. The atomic number of carbon is 6, and the atomic number of copper is 29. What does that tell you about carbon and copper atoms? Every carbon atom has six protons, and every copper atom has twenty-nine protons.

The atomic number also tells you something about the number of electrons in an atom. Remember that protons are positively charged and electrons are negatively charged. Because there is no overall charge on an atom, the number of positive charges must be balanced by the number of negative charges. In other words, the number of electrons is the same as the number of protons. If you know the atomic number of an element, you also know the number of electrons in the atoms of that element as well.

atomic mass	page 14
atomic number	page 14
element	page 14
isotopes	page 16
mass number	page 16

Properties of Subatomic Particles

Particle	Symbol	Charge	Mass (g)	Location
Electron	e^-	−1	9.109×10^{-28}	Outside the nucleus
Proton	p^+	+1	1.673×10^{-24}	Inside the nucleus
Neutron	n^o	0	1.675×10^{-24}	Inside the nucleus

▲ The charges of the proton and electron are equal and opposite; the neutron has no electric charge. The masses of the neutron and proton are about the same. The mass of the electron is considerably less than that of either the proton or neutron.

Mass Number

The **mass number** is the sum of the protons and neutrons in an atom. All atoms of the same element have the same number of protons. But they can have a different number of neutrons. Look at the three forms of hydrogen in the table below. To find the mass number of each, add the number of protons to the number of neutrons. You can now compare the mass numbers.

Isotopes

Atoms that have the same number of protons but different numbers of neutrons are called **isotopes**. The three forms of hydrogen in the chart below are isotopes. Generally, isotopes of an element have similar properties. However, some isotopes are unstable. An unstable atom is one that changes into another element by giving off particles and energy. This type of isotope is called radioactive. Tritium is a radioactive isotope of hydrogen.

Isotopes are identified by their mass numbers. In one method of naming isotopes, the name or symbol of the element is followed by the mass number with a hyphen in between. For example, tritium is written as hydrogen-3, or H-3. The uranium isotope is written as uranium-235, or U-235.

Mass Numbers of Hydrogen Atoms

Form of Hydrogen	Number of Protons	Atomic Number	Number of Neutrons	Mass Number
Protium	1	1	0	1 + 0 = 1
Deuterium	1	1	1	1 + 1 = 2
Tritium	1	1	2	1 + 2 = 3

◀ Each hydrogen atom has one proton in the nucleus. However, the three forms have different mass numbers because they have different numbers of neutrons.

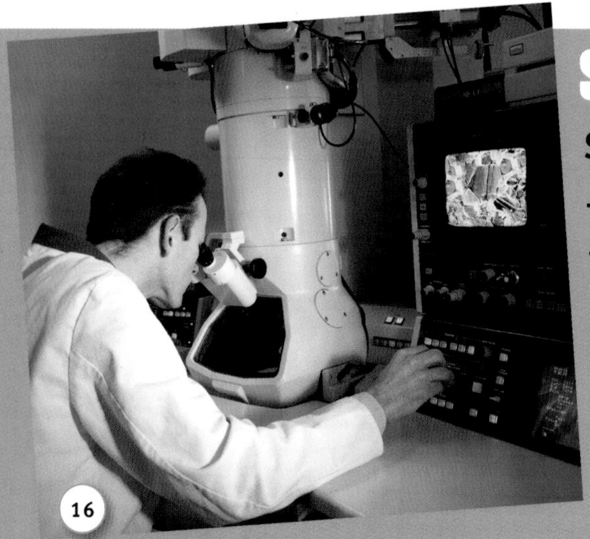

Science and Technology

Scanning Transmission Electron Microscopes

To look at extremely small objects, scientists use scanning transmission electron microscopes (STEM). Electron microscopes use a beam of electrons in much the same way that light microscopes use light. STEMs provide greater magnification and more control than light microscopes do. Developed in the 1960s, the STEM is one of the most useful instruments in scientific research today.

Several Naturally Occurring Isotopes

Isotope	Atomic Number	Mass Number	Percentage Natural Abundance	Atomic Mass Unit (amu)
Hydrogen-1	1	1	99.98	1.00783
Hydrogen-2	1	2	0.02	2.01410
Carbon-12	6	12	98.91	12.000
Carbon-13	6	13	1.09	13.0034
Oxygen-16	8	16	99.76	15.9499
Oxygen-17	8	17	0.037	16.9991
Oxygen-18	8	18	0.200	17

◀ Not all isotopes of an element exist in the same amounts. More than 99% of all the hydrogen atoms found on Earth are Protium (hydrogen-1).

The name of an isotope can be used to determine the number of neutrons in its atoms. Remember that the atomic number in all atoms of an element is the same. If the atomic number is subtracted from the mass number, the result is the number of neutrons. For example, uranium-235 has an atomic number of 92, so this isotope has 235 – 92 = 143 neutrons.

1P
Protium (Hydrogen-1)

1P 1N
Deuterium (Hydrogen-2)

1P 2N
Tritium (Hydrogen-3)

▲ Two of hydrogen's isotopes are stable. The third is unstable and radioactive.

Atomic Mass

In nature, most elements exist as a mixture of two or more stable isotopes. Each isotope has a different mass and occurs in a different amount, or abundance. To describe the mass of the mixture of isotopes, scientists use atomic mass. The atomic mass of an element is an average of the masses of the naturally occurring isotopes. A weighted average takes into account the abundance in which the isotopes naturally occur. Atomic mass is measured in atomic mass units (amu). One atomic mass unit is exactly 1/12 the mass of a carbon-12 atom. The atomic mass of an element is often a clue to the abundance of its isotopes. For example, the atomic mass of carbon is 12.01, a number closer to the mass of carbon-12 than carbon-13. Thus carbon-12 must be more abundant.

Science and Math

Averages

An average is a measure of the most typical value in a set of data. An average is calculated by adding the values and then dividing by the number of values. Another term for an average is a mean. Here's how to calculate the average of the values below:

6, 10, 18, 12, 8, 15, 20, 7

$$\text{Average} = \frac{6 + 10 + 18 + 12 + 8 + 15 + 20 + 7}{8} = 12$$

Hands-On Science

Modeling Isotopes

A model is a diagram, image, or three-dimensional representation of something. Using models is one way scientists can show and share information. Working with models can also help scientists better understand an object or process. You can use models to help understand isotopes.

TIME: 45 minutes

MATERIALS:
- round items of different colors, such as gumdrops or marbles
- glue or frosting
- mass balance

ARRANGEMENT: pairs of students

▲ C-12

▲ C-13

▲ C-14

Proton

Neutron

Electron

Procedure

STEP 1: Make a table similar to the one below.

Form of Carbon	Number of Protons	Number of Neutrons	Mass of Nucleus

STEP 2: Carbon has three isotopes: C-12, C-13, and C-14. Complete the table you made for each isotope of carbon.

STEP 3: Look at the isotopes of carbon. Use the materials you have to make a model of the nucleus of each isotope of carbon.

STEP 4: Carefully measure the mass of each model. Record your measurements in the table.

STEP 5: How are the models alike? How are they different?

Science to Science

Geology and Physical Science

Carbon-14 dating is a way of using an isotope of carbon to determine the ages of ancient objects. Radioactive carbon-14 changes into an isotope of nitrogen over time. Scientists can measure the amounts of these different isotopes in a sample to figure out the sample's age. Carbon-14 dating can be used on objects up to 50,000 years old. It is often used to find the ages of bones, teeth, and cloth.

Summing Up

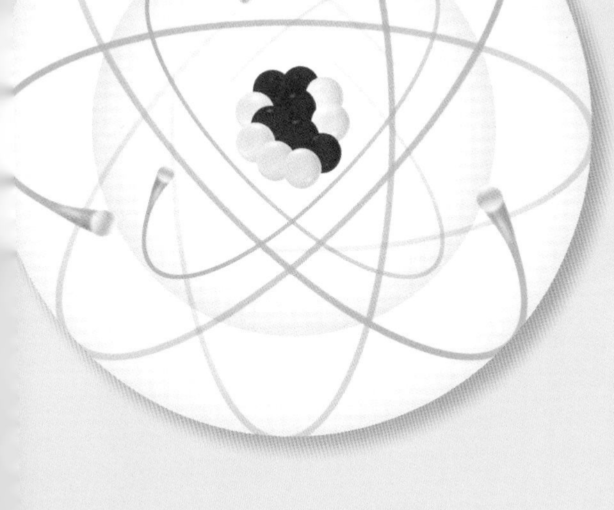

Atoms can be described by their numbers of protons, neutrons, and electrons.

The atomic number is the number of protons in the nucleus of an atom.

For a neutral atom, the number of protons is equal to the number of electrons.

The sum of the protons and neutrons is the mass number of an atom.

Atoms of the same element with different mass numbers are called isotopes.

The atomic mass of an element is determined from the weighted average of the masses of its isotopes.

Putting it All Together

Choose from the following research activities. Work independently, in pairs, or in a small group. Share your responses with the class. Listen to other groups present their answers.

1. Describe each subatomic particle, including its mass, charge, and location. Then make a simple model of an atom using materials from around the classroom or your home.

2. Explain how the atomic number is unique for every element, but the mass number is not. Use examples to create a poster that illustrates your explanation.

3. Carbon has fifteen known isotopes. One of those isotopes was chosen as the basis for atomic weights. Conduct research to find out about the isotopes of carbon. Prepare a poster explaining what an isotope is and why carbon was chosen for this standard.

Matter of Artistic Interpretation

Cartoonist's Notebook • Illustrated by Kylee Solari

PART OF BEING A SCIENTIST IS THINKING CREATIVELY. WHAT ARE SOME WAYS THAT THE SCIENTISTS IN THIS BOOK HAD TO THINK CREATIVELY IN ORDER TO SOLVE PROBLEMS?

Chapter 3

The Periodic Table

How are chemical elements organized in the periodic table?

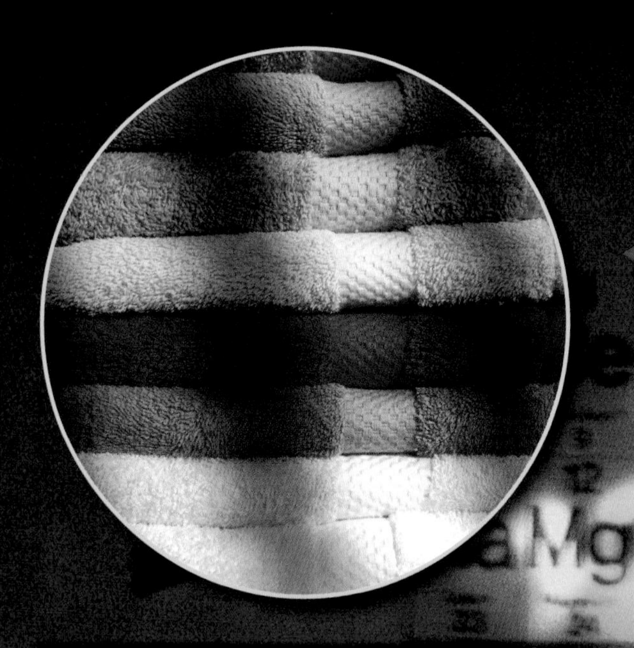

These towels are organized by color, size, and how they are used. In a similar way, chemists organize chemical elements by their physical and chemical properties.

Everyday Science

Tables

The world has millions of different organisms. In order to better understand and describe them, scientists rely on taxonomy. Taxonomy is the science of naming, describing, and classifying living things into groups. The largest groups are known as domains. Each domain is divided into kingdoms. Organisms within the same kingdom are further divided by phylum, class, order, family, genus, and species. We also use tables to classify elements.

Essential Vocabulary

group	page 27
metalloids	page 31
metals	page 30
nonmetals	page 31
periodic law	page 27
periods	page 27
valence electrons	page 33

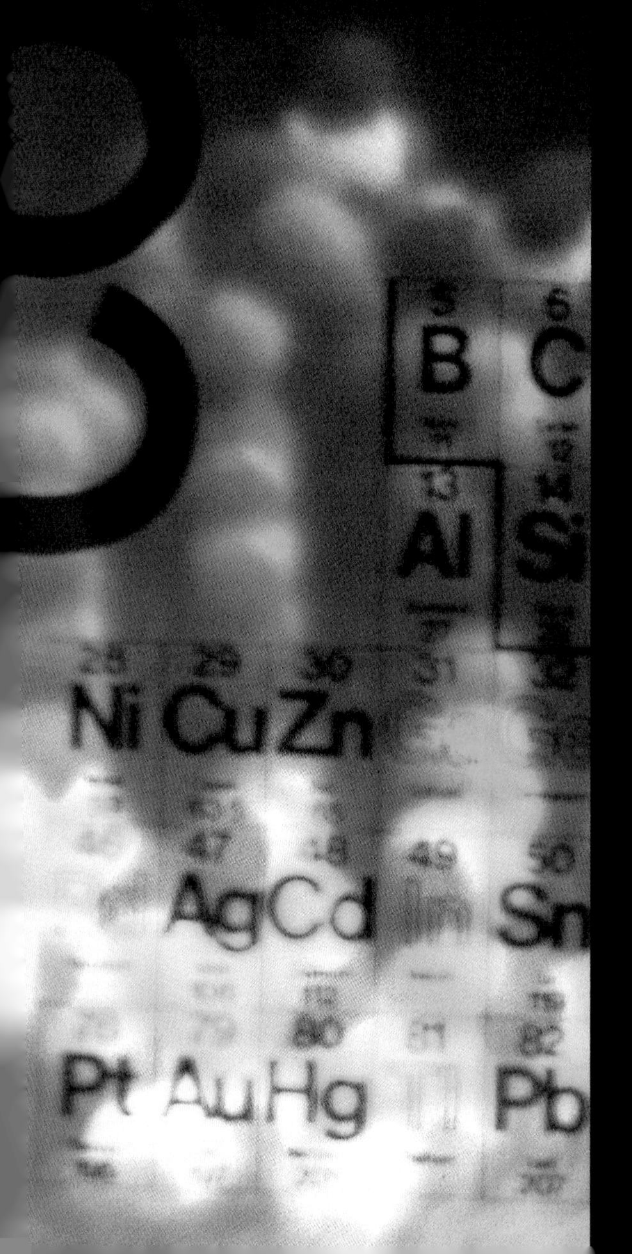

By 1860, chemists had discovered about sixty different elements. Through investigations, the chemists were learning about the properties of the elements. This task was slowed by the fact that there was no accurate way to determine an element's atomic mass. Because different chemists used different atomic masses for the same elements, a variety of descriptions for each element resulted. Chemists could not agree on their results.

In September of that year, standard values for atomic mass were established. Now chemists could focus on finding relationships between atomic mass and the properties of the elements. Their goal was to organize elements in much the same way you might organize your clothes or your music—by the characteristics they have in common.

MENDELEEV'S Contributions

At about this same time, the Russian chemist Dmitri Mendeleev (dih-MEE-tree men-deh-LAY-ef) was trying to organize the elements by their properties. Mendeleev had written the names and properties of each known element on a separate card.

Mendeleev tried many different arrangements of the cards. Eventually, he discovered that when he arranged the element cards in a table based on atomic masses, a pattern emerged. Elements with similar properties appeared at regular intervals in vertical columns of the table.

For the pattern to work, Mendeleev had to leave three blank spaces in the table. He predicted that new elements would be discovered to fill those blanks—and he was right! By 1886, all three elements were discovered. The properties of these elements were very similar to those he predicted. The success of predicting these elements persuaded most chemists to accept his table.

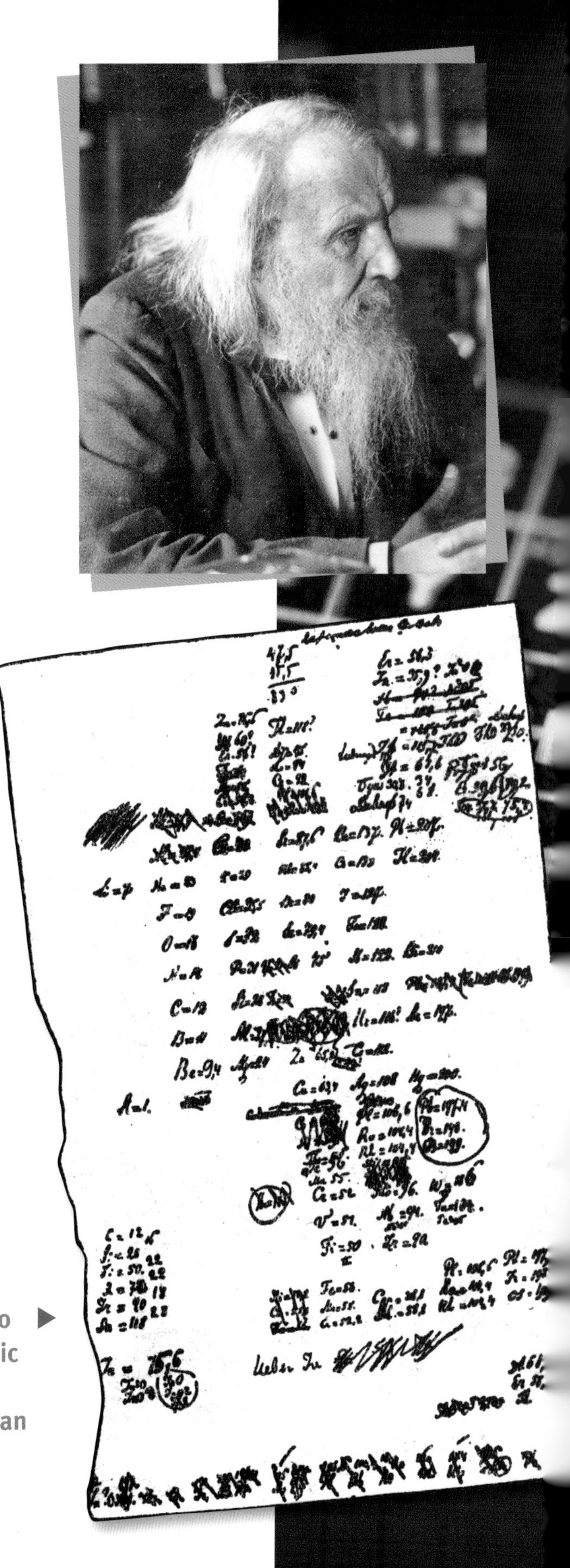

Mendeleev organized elements into vertical columns according to atomic mass. To honor his contribution, element 101 was named Mendelevian when it was produced in 1955.

— 70 —

но въ ней, мнѣ кажется, уже ясно выражается примѣнимость вы ставляемаго мною начала ко всей совокупности элементовъ, пай которыхъ извѣстенъ съ достовѣрностію. На этотъ разъ я и желалъ преимущественно найдти общую систему элементовъ. Вотъ этотъ опытъ:

			Ti=50	Zr=90	?=180.
			V=51	Nb=94	Ta=182.
			Cr=52	Mo=96	W=186.
			Mn=55	Rh=104,4	Pt=197,4
			Fe=56	Ru=104,4	Ir=198.
			Ni=Co=59	Pl=106,6	Os=199.
H=1			Cu=63,4	Ag=108	Hg=200.
	Be=9,4	Mg=24	Zn=65,2	Cd=112	
	B=11	Al=27,4	?=68	Ur=116	Au=197?
	C=12	Si=28	?=70	Sn=118	
	N=14	P=31	As=75	Sb=122	Bi=210
	O=16	S=32	Se=79,4	Te=128?	
	F=19	Cl=35,5	Br=80	I=127	
Li=7	Na=23	K=39	Rb=85,4	Cs=133	Tl=204
		Ca=40	Sr=87,6	Ba=137	Pb=207.
		?=45	Ce=92		
		?Er=56	La=94		
		?Yt=60	Di=95		
		?In=75,6	Th=118?		

а потому приходится въ разныхъ рядахъ имѣть различное измѣненіе разностей, чего нѣтъ въ главныхъ числахъ предлагаемой таблицы. Или же придется предполагать при составленіи системы очень много недостающихъ членовъ. То и другое мало выгод... ...ается притомъ, наиболѣе естественнымъ составить кубическую сис... ванія не повели... казать то разн... начала, высказ...

ПЕРИОДИЧЕСКАЯ СИСТЕМА ЭЛЕМЕНТОВ

Периоды	Ряды	I	II	III	IV	V	VI	VII	VIII			0
		ГРУППЫ ЭЛЕМЕНТОВ										
1	I	H 1 1,008										He 2 4,003
2	II	Li 3 6,940	Be 4 9,02	5 B 10,82	6 C 12,010	7 N 14,008	8 O 16,000	9 F 19,00				Ne 10 20,183
3	III	Na 11 22,997	Mg 12 24,32	13 Al 26,97	14 Si 28,06	15 P 30,98	16 S 32,06	17 Cl 35,457				Ar 18 39,944
4	IV	K 19 39,096	Ca 20 40,08	Sc 21 45,10	Ti 22 47,90	V 23 50,95	Cr 24 52,01	Mn 25 54,93	Fe 26 55,85	Co 27 58,94	Ni 28 58,69	
	V	29 Cu 63,57	30 Zn 65,38	31 Ga 69,72	32 Ge 72,60	33 As 74,91	34 Se 78,96	35 Br 79,916				Kr 36 83,7
5	VI	Rb 37 85,48	Sr 38 87,63	Y 39 88,92	Zr 40 91,22	Nb 41 92,91	Mo 42 95,95	Ma 43 —	Ru 44 101,7	Rh 45 102,91	Pd 46 106,7	
	VII	47 Ag 107,88	48 Cd 112,41	49 In 114,76	50 Sn 118,70	51 Sb 121,76	52 Te 127,61	53 J 126,92				Xe 54 131,3
6	VIII	Cs 55 132,91	Ba 56 137,36	La 57 * 138,92	Hf 72 178,6	Ta 73 180,88	W 74 183,92	Re 75 186,31	Os 76 190,2	Ir 77 193,1	Pt 78 195,23	
	IX	79 Au 197,2	80 Hg 200,61	81 Tl 204,39	82 Pb 207,21	83 Bi 209,00	84 Po 210	85 —				Rn 86 222
7	X	87 —	Ra 88 226,05	Ac 89 227	Th 90 232,12	Pa 91 231	U 92 238,07					

* ЛАНТАНИДЫ 58-71

Ce 58 140,13	Pr 59 140,92	Nd 60 144,27	61 —	Sm 62 150,43	Eu 63 152,0	Gd 64 156,9
Tb 65 159,2	Dy 66 162,46	Ho 67 164,94	Er 68 167,2	Tu 69 169,4	Yb 70 173,04	Cp 71 174,99

They Made a Difference

Glenn Seaborg

The last major change to the periodic table resulted from the work of Glenn Seaborg. Seaborg is best known for discovering the element plutonium (atomic number 94) and the elements that followed it up to atomic number 102. He and his colleagues are responsible for discovering more than 100 isotopes. For his work, he shared in the 1951 Nobel Prize in Chemistry and in 1997 had the element with atomic number 106 named seaborgium in his honor.

MOSELEY'S Improvements

Although Mendeleev's table was generally successful at organizing the elements, it did have some shortcomings. Most important, Mendeleev had to rearrange some elements so that they were no longer in order of increasing atomic mass. For example, the atomic mass of iodine (127) is lower than that of tellurium (128), so iodine should come before tellurium in the table. Iodine chemically behaved more like the elements fluorine, chlorine, and bromine. Tellurium chemically behaved more like oxygen, sulfur, and selenium. Because of this observation, Mendeleev switched their positions.

In 1913, the British chemist Henry Moseley subjected the known elements to X-rays. In doing so, he was able to determine the atomic number of each element. He recognized that he could correct Mendeleev's table by arranging the elements in order of increasing atomic number instead of atomic mass. Now the arrangement of iodine and tellurium made sense. Iodine has a higher atomic number than tellurium. Therefore, iodine should come after tellurium in the table. Today, the periodic table is constructed to show Moseley's correction.

Henry Moseley

Periodic Law

Today the table of elements is known as the periodic table. A pattern that repeats at regular intervals is called periodic. The days of the week are periodic because the pattern recurs every seven days. The hands of a clock pass each hour at regular, periodic intervals. The ocean tides rise and fall each day in a predictable, periodic pattern. The phases of the moon change in a periodic pattern throughout the month.

The properties of the elements are also periodic. The **periodic law** states that the chemical and physical properties of elements repeat in a pattern when the elements are arranged in order of increasing atomic number. For this reason, the table of elements is known as the periodic table.

The modern periodic table looks very different from Mendeleev's version. At the present time, scientists have identified more than 100 elements. The periodic table is shown on the following two pages. The elements are arranged in horizontal rows called **periods**. Each vertical column is known as a **group**, or family. Both periods and groups are identified by specific numbers.

checkpoint

Make Connections

Look for other examples of things that are periodic. Consider magazines, television shows, animal and plant behaviors, and changes that occur in nature. Describe your examples to the class.

SUN	MON	TUE	WED	THU	FRI	SAT
27	28	29	30	31	1	2
3	4	5	6	7	8	9
10	11	12	13	14	15	16
17	18	19	20	21	22	23
24 / 31	25	26	27	28	29	30

The passing from night to day occurs in a periodic way. In a similar way, the properties of the elements repeat when they are arranged in the periodic table.

Periodic Table of the Elements

- Alkali Metals
- Alkaline Earth Metals
- Transition Metals
- Lanthanoids
- Actinoids
- Other Metals
- Metalloids
- Nonmetals
- Noble Gases

Atomic Mass
symbol
Name
Atomic #

Physical states are at normal temperature and pressure.

Period	1	2	3	4	5	6	7	8	9
1	1.00794 H Hydrogen 1								
2	6.941 Li Lithium 3	9.012182 Be Beryllium 4							
3	22.98976928 Na Sodium 11	24.3050 Mg Magnesium 12							
4	39.0983 K Potassium 19	40.078 Ca Calcium 20	44.955912 Sc Scandium 21	47.867 Ti Titanium 22	50.9415 V Vanadium 23	51.9961 Cr Chromium 24	54.938045 Mn Manganese 25	55.845 Fe Iron 26	58.933195 Co Cobalt 27
5	85.4678 Rb Rubidium 37	87.62 Sr Strontium 38	88.90585 Y Yttrium 39	91.224 Zr Zirconium 40	92.90638 Nb Niobium 41	95.96 Mo Molybdenum 42	(97.9072) Tc Technetium 43	101.07 Ru Ruthenium 44	102.90550 Rh Rhodium 45
6	132.9054519 Cs Caesium 55	137.327 Ba Barium 56	57 - 71	178.49 Hf Hafnium 72	180.94788 Ta Tantalum 73	183.84 W Tungsten 74	186.207 Re Rhenium 75	190.23 Os Osmium 76	192.217 Ir Iridium 7
7	(223) Fr Francium 87	(226) Ra Radium 88	89 - 103	(261) Rf Rutherfordium 104	(262) Db Dubnium 105	(266) Sg Seaborgium 106	(264) Bh Bohrium 107	(277) Hs Hassium 108	(268) Mt Meitnerium 10

138.90547 La Lanthanum 57	140.116 Ce Cerium 58	140.90765 Pr Praseodymium 59	144.242 Nd Neodymium 60	(145) Pm Promethium 61	150.36 Sm Samarium 62	151.964 Eu Europium 63	157.25 Gd Gadolinium 6
(227) Ac Actinium 89	232.03806 Th Thorium 90	231.03588 Pa Protactinium 91	238.02891 U Uranium 92	(237) Np Neptunium 93	(244) Pu Plutonium 94	(243) Am Americium 95	(247) Cm Curium 9

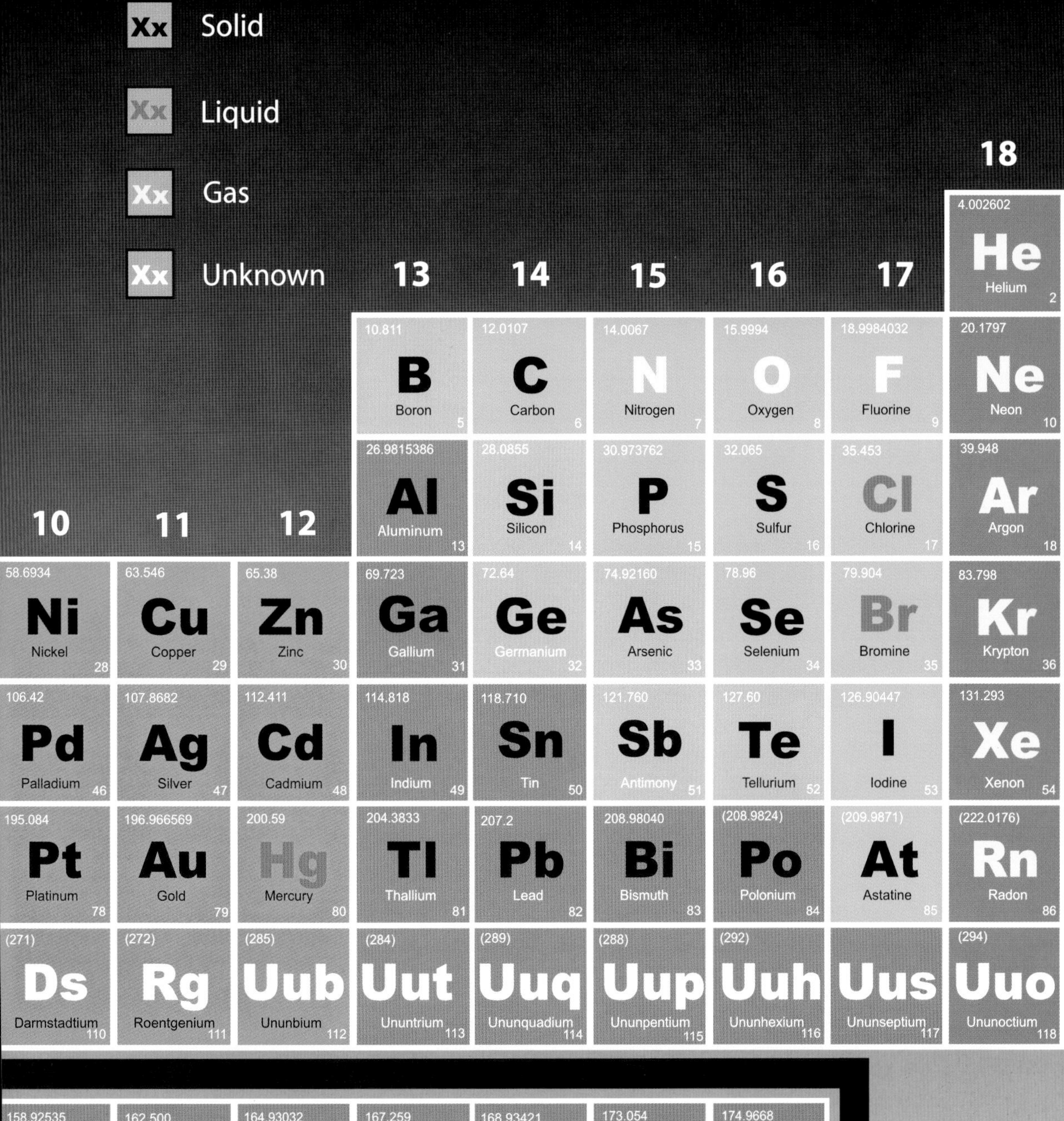

Key	
Xx	Solid
Xx	Liquid
Xx	Gas
Xx	Unknown

10	11	12	13	14	15	16	17	18
								4.002602 He Helium 2
			10.811 B Boron 5	12.0107 C Carbon 6	14.0067 N Nitrogen 7	15.9994 O Oxygen 8	18.9984032 F Fluorine 9	20.1797 Ne Neon 10
			26.9815386 Al Aluminum 13	28.0855 Si Silicon 14	30.973762 P Phosphorus 15	32.065 S Sulfur 16	35.453 Cl Chlorine 17	39.948 Ar Argon 18
58.6934 Ni Nickel 28	63.546 Cu Copper 29	65.38 Zn Zinc 30	69.723 Ga Gallium 31	72.64 Ge Germanium 32	74.92160 As Arsenic 33	78.96 Se Selenium 34	79.904 Br Bromine 35	83.798 Kr Krypton 36
106.42 Pd Palladium 46	107.8682 Ag Silver 47	112.411 Cd Cadmium 48	114.818 In Indium 49	118.710 Sn Tin 50	121.760 Sb Antimony 51	127.60 Te Tellurium 52	126.90447 I Iodine 53	131.293 Xe Xenon 54
195.084 Pt Platinum 78	196.966569 Au Gold 79	200.59 Hg Mercury 80	204.3833 Tl Thallium 81	207.2 Pb Lead 82	208.98040 Bi Bismuth 83	(208.9824) Po Polonium 84	(209.9871) At Astatine 85	(222.0176) Rn Radon 86
(271) Ds Darmstadtium 110	(272) Rg Roentgenium 111	(285) Uub Ununbium 112	(284) Uut Ununtrium 113	(289) Uuq Ununquadium 114	(288) Uup Ununpentium 115	(292) Uuh Ununhexium 116	(294) Uus Ununseptium 117	(294) Uuo Ununoctium 118

158.92535 Tb Terbium 65	162.500 Dy Dysprosium 66	164.93032 Ho Holmium 67	167.259 Er Erbium 68	168.93421 Tm Thulium 69	173.054 Yb Ytterbium 70	174.9668 Lu Lutetium 71
(247) Bk Berkelium 97	(251) Cf Californium 98	(252) Es Einsteinium 99	(257) Fm Fermium 100	(258) Md Mendelevium 101	(259) No Nobelium 102	(262) Lr Lawrencium 103

Classes of Elements

The elements of the periodic table are classified into three distinct groups: metals, nonmetals, and metalloids. Look back at the table on the previous pages. You can use the color key at the top to find elements in each category.

▼ Metals are often recognized because they are shiny. They reflect the light that strikes them.

Metals

Most of the elements in the periodic table are **metals**. You are probably familiar with some metals, such as gold, silver, aluminum, and copper. Most metals have similar properties. For example, they tend to be shiny. Metals are generally good conductors of heat and electricity. Metals are malleable, which means that they can be flattened without breaking. They are also ductile, which means they can be drawn into thin wires. With the exception of mercury, metals are solids at room temperature. They melt only at high temperatures.

▼ Many metals can be drawn into thin wires. Stainless steel, which is a mixture of iron and chromium, is used for braces that straighten teeth.

▲ Because they are good conductors, metal wires are used in electric circuits.

▲ Gold is the most malleable substance known. It can even be pressed into flakes thin enough to be sprinkled on foods.

Nonmetals

With the exception of hydrogen, all of the **nonmetals** appear on the right side of the periodic table. The properties of nonmetals are essentially the opposite of those of metals. Nonmetals are not shiny. They are poor conductors of heat and electricity, and they are neither malleable nor ductile. Unlike metals, nonmetals break or shatter when they are hit. In addition, nonmetals exist in all three states of matter at room temperature.

Metalloids

Notice the zigzag line toward the right side of the periodic table. These are the **metalloids**. These elements have some properties of metals and some properties of nonmetals. The metalloids include boron, silicon, germanium, arsenic, antimony, and tellurium. Another name for metalloids is semimetals. Metalloids can be shiny or dull. They generally conduct heat and electricity better than nonmetals, but not as well as metals.

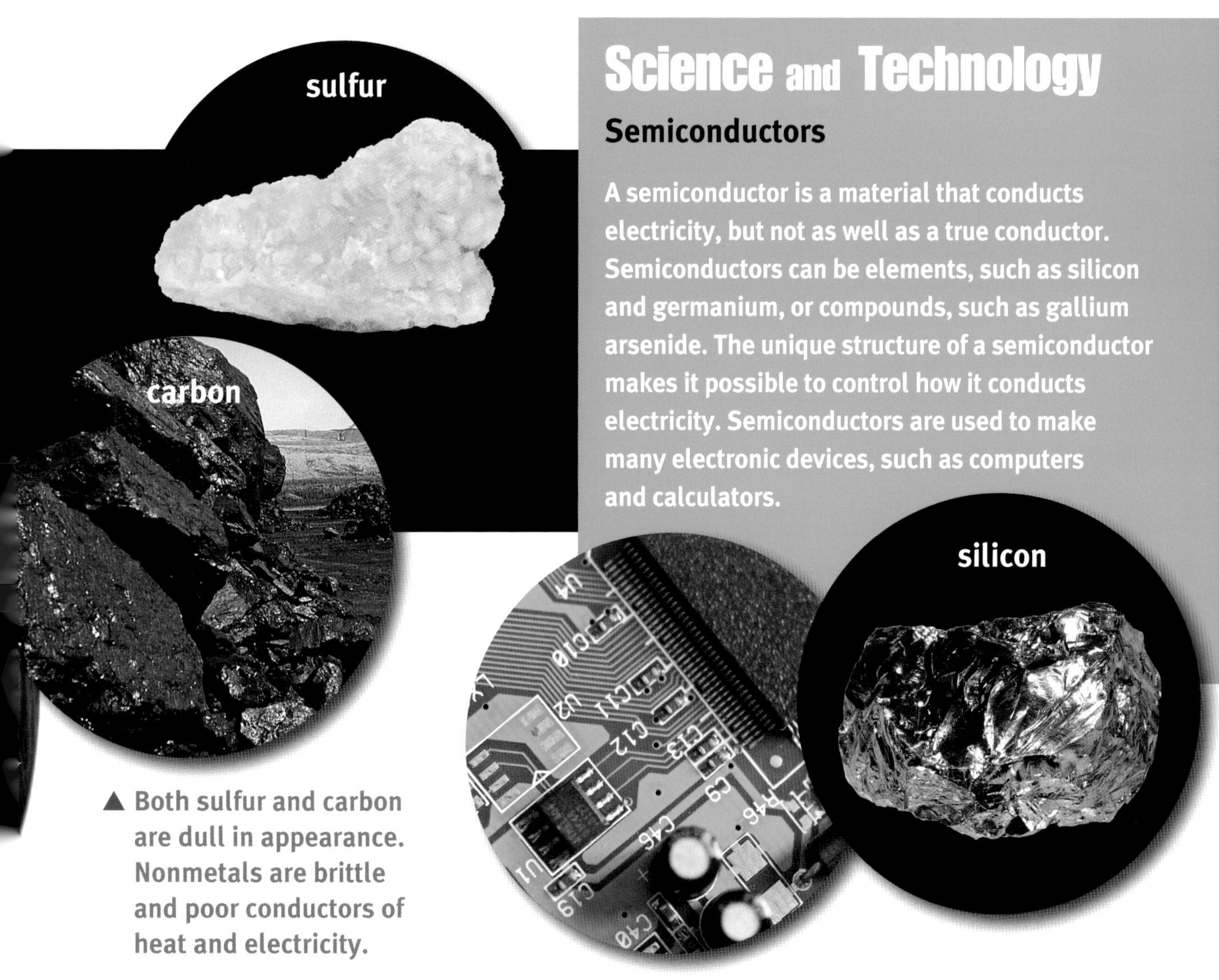

Science and Technology

Semiconductors

A semiconductor is a material that conducts electricity, but not as well as a true conductor. Semiconductors can be elements, such as silicon and germanium, or compounds, such as gallium arsenide. The unique structure of a semiconductor makes it possible to control how it conducts electricity. Semiconductors are used to make many electronic devices, such as computers and calculators.

▲ Both sulfur and carbon are dull in appearance. Nonmetals are brittle and poor conductors of heat and electricity.

▲ Pure silicon is a metalloid used to make computer chips. Silicon is shiny like a metal, but it is a poor conductor like a nonmetal.

Hands-On Science

Testing for Conductivity

Conductivity is the ability to conduct energy. Most metals are good conductors. You can use a conductivity test to determine whether an object is a metal or a nonmetal. Here's how:

TIME: 45 minutes

MATERIALS:
- C or D battery
- small bulb (1.5 or 3 volts)
- wire
- tape
- objects for testing (paper, penny, paper clip, aluminum foil, cork, wooden stick)

ARRANGEMENT: 3–4 students

Procedure

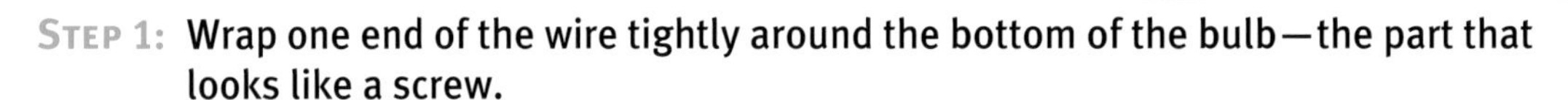

STEP 1: Wrap one end of the wire tightly around the bottom of the bulb—the part that looks like a screw.

STEP 2: Tape the other end of the wire to the negative terminal (flat end) of the battery.

STEP 3: Make a table similar to the table below.

Object	Prediction	Bulb On	Bulb Off

STEP 4: Select an object to test. Predict whether you think it will or will not conduct electricity. Write "Yes" or "No" in the table to correspond with your prediction.

STEP 5: Place the object near the battery. Press the positive terminal of the battery (+) to one side of the object. Press the metal end of the bulb to the other end of the object. Observe the bulb. Check the appropriate box in the table to record your observation. Be careful, as the bulb, and other objects may grow hot during test.

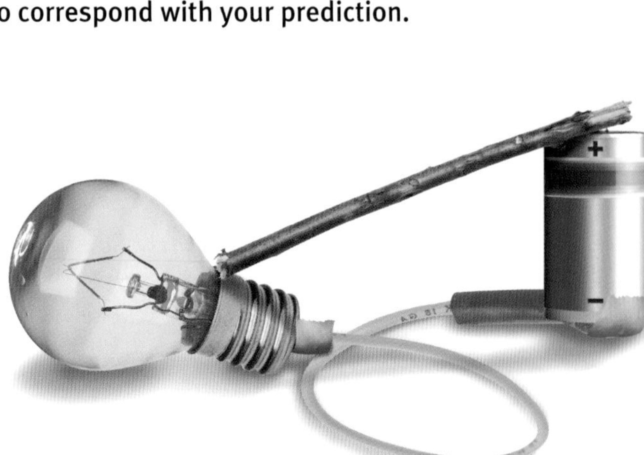

STEP 6: Repeat steps 4 and 5 for the other objects you are testing.

STEP 7: Based on your observations, classify the test objects as metals or other materials. If an object that you know is a metal did not cause the bulb to light, propose a reason why. For example, the metal might be coated with plastic.

Groups of Elements

The elements in each group of the periodic table have something in common. Recall that electrons travel in different energy levels around the nucleus. The electrons in the outermost level are known as **valence electrons**. Atoms are most stable when they have a complete set of valence electrons. For most atoms, a complete set of valence electrons consists of eight electrons. Atoms give up, gain, or share electrons to become stable (to complete their outermost level). All of the elements in a group have the same number of valence electrons. As a result, they behave in the same way in order to become stable.

GROUP 1

Alkali Metals

The alkali metals are soft and have low densities. Their atoms have one valence electron. To become stable, they give this electron away rather easily, so these elements are said to be very reactive. Alkali metals are so reactive, they are always found combined with other elements.

Alkali metals react violently with water. They are usually stored in oil to prevent them from reacting with water or oxygen in the air.

GROUP 2

Alkaline-Earth Metals

The alkaline-earth metals have two valence electrons. Because it is more difficult to give away two electrons than one electron, these metals are not quite as reactive as the alkali metals. However, they are still very reactive, and are always found combined with other elements.

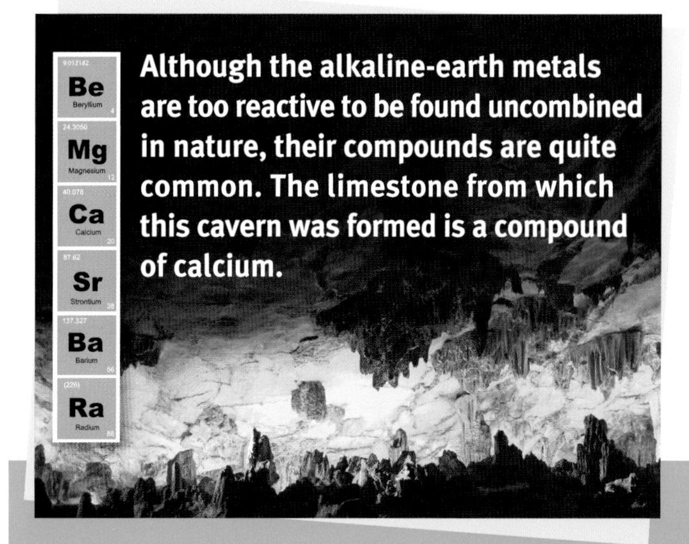

Although the alkaline-earth metals are too reactive to be found uncombined in nature, their compounds are quite common. The limestone from which this cavern was formed is a compound of calcium.

Science to Science

Chemistry and Nutrition

Did you have your metals today? Two alkaline-earth metals are an essential part of your diet. Calcium is necessary for your nerves to function properly, your blood to clot, and your muscles to contract. It is also an important component of bones and teeth. Magnesium supports a healthy immune system, keeps the heart rhythm steady, and keeps bones strong. Calcium and magnesium are found in dairy products, legumes, nuts, whole grains, and vegetables.

GROUPS 3–12

Transition Metals

Groups 3 through 12 of the periodic table do not have individual names. Instead, they are described as one group—the transition metals. The transition metals have properties that are similar to other metals, but they do not fit into other groups.

Some transition metals are shown below the main part of the periodic table in order to keep the table from becoming too wide. These elements are the lanthanoids and actinoids. Compounds of the lanthanoids are used to produce lamps, lasers, movie projectors, and petroleum products. All of the actinoids are radioactive. Elements with atomic numbers greater than 94 (plutonium) do not occur in nature. They are produced in laboratories.

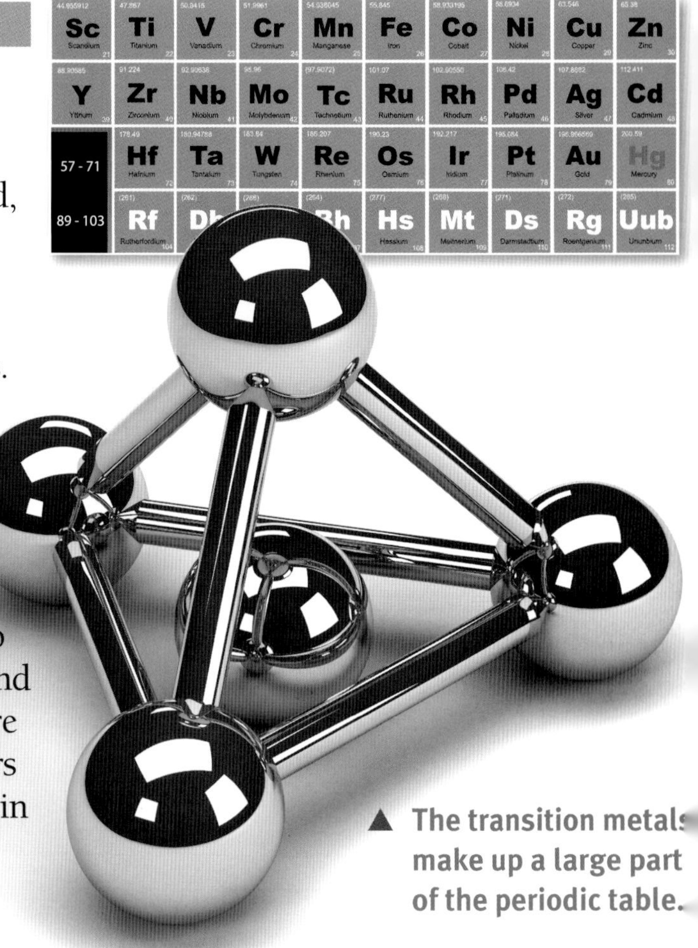

▲ The transition metals make up a large part of the periodic table.

Career

Metallurgist

A metallurgist studies the properties and uses of metals and mixtures of metals, such as steel. A chemical metallurgist is involved with removing metals from ores. A physical metallurgist studies the behavior of metals under stress and changes in temperature. Process metallurgists investigate how metals can be shaped and joined together. Daily work for any metallurgist might involve research and development, design and manufacturing, or quality assurance.

Metallurgists generally require a bachelor's degree along with a master's degree in science, technology, or engineering.

GROUP 13

Boron Group

This group contains one metalloid and four metals. Boron is a gray powder, whereas all the other members of this group are soft, silvery metals. All of the elements in this group have three valence electrons. When they react with other elements, they lose these electrons. One of the most important elements in this group is aluminum. It is the third most abundant element in Earth's crust.

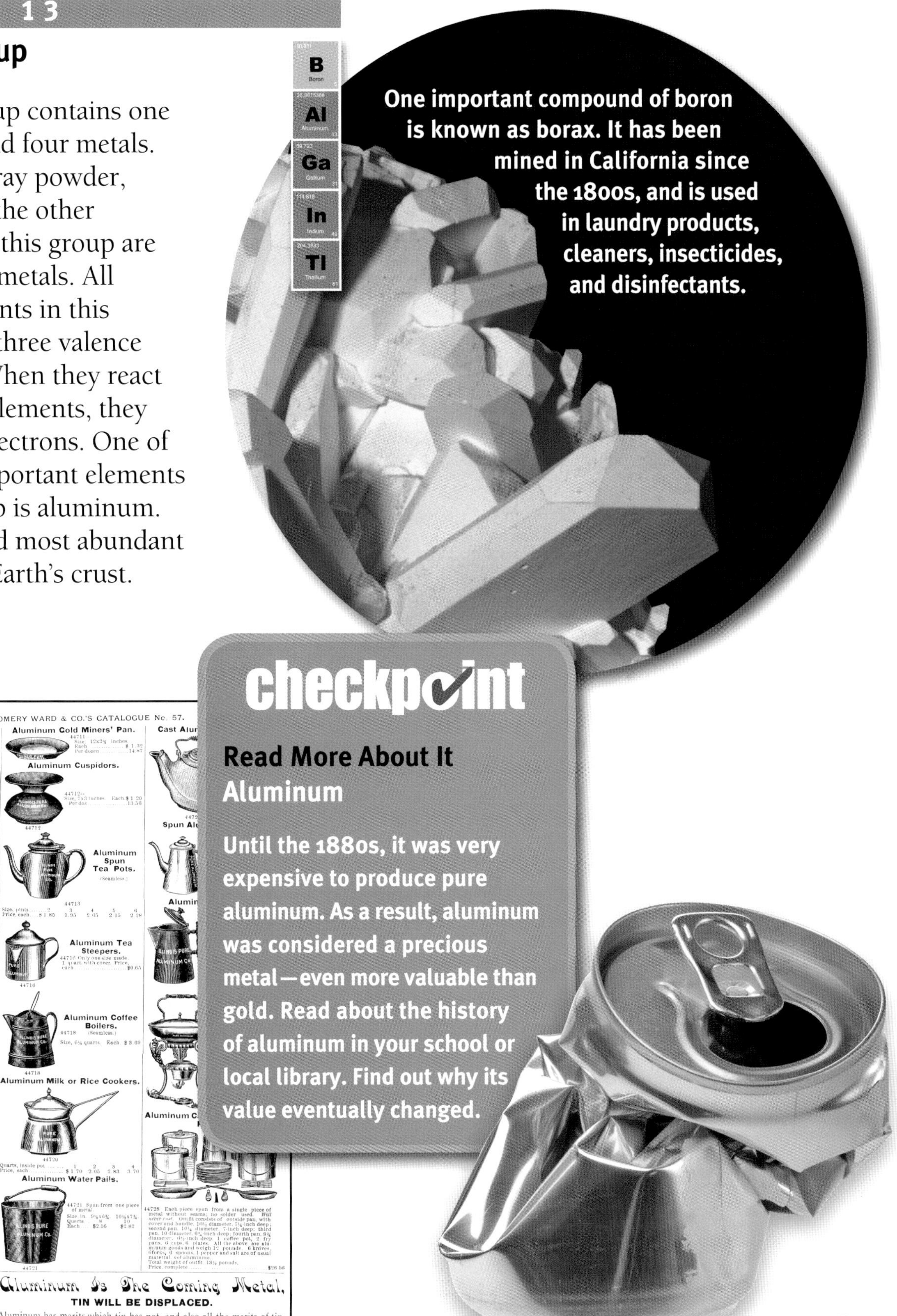

One important compound of boron is known as borax. It has been mined in California since the 1800s, and is used in laundry products, cleaners, insecticides, and disinfectants.

checkpoint

Read More About It
Aluminum

Until the 1880s, it was very expensive to produce pure aluminum. As a result, aluminum was considered a precious metal—even more valuable than gold. Read about the history of aluminum in your school or local library. Find out why its value eventually changed.

Group 14

Carbon Group

The carbon group contains a nonmetal, two metalloids, and two metals. Each element in this group has four valence electrons. The elements generally share (rather than lose or gain) electrons with other elements to become stable.

One of the most interesting elements in this group is carbon itself. Carbon is one of the most abundant elements in the universe. It can form more compounds than any other element except hydrogen. Plants use carbon dioxide, a compound of carbon, to make food. Fossil fuels (coal, oil, and natural gas) are made up of pure carbon or carbon compounds.

Silicon is the second most abundant element in Earth's crust. Silicon exists in a compound called silicon dioxide, or silica. One form of silica is quartz, which makes up beach sand. Compounds of silicon are also found in gemstones such as emeralds

A diamond, which is very hard, ▶ and graphite, which is very soft, are two forms of carbon.

▼ Sandy beaches and sparkling gems are made up of compounds of silicon

Science to Science

Physical Science and Life Science

Have you ever heard the term "carbon-based life form"? You are such a life form! Compounds that contain carbon are found in all known living things. Carbon, along with a few other elements, forms many of the compounds essential to life, such as carbohydrates, proteins, lipids, and nucleic acids. Carbon is exchanged between living things and their environment through the carbon cycle.

GROUP 15

Nitrogen Group

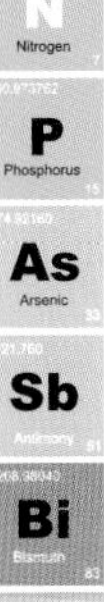

The nitrogen group consists of two nonmetals, two metalloids, and only one metal. All of the elements in this group have five valence electrons. They need to gain three electrons to become stable.

One of the most important elements in his group is nitrogen. About 78 percent of the atmosphere is made up of nitrogen gas. Nitrogen is colorless, odorless, and nonflammable. Proteins, which are nitrogen-containing compounds, are essential to living things. Nitrogen flows through the environment in the nitrogen cycle.

Compounds of the element phosphorus are used in fertilizers. Phosphorus is used to make cleaning agents, water softeners, and in the production of steel. In living things, phosphorus is found in the nucleic acids that determine the traits of an organism.

ne use of nitrogen ▶ in explosives.

GROUP 16

Oxygen Group

O Oxygen
S Sulfur
Se Selenium
Te Tellurium
Po Polonium

The oxygen group contains three nonmetals, one metalloid, and one metal. Each element in this group has six valence electrons. These elements combine with other elements to gain two electrons.

Oxygen is a colorless and odorless gas that makes up about 21 percent of air. Living things need a continuous supply of oxygen because it is used to release the energy stored in food. Oxygen combines with other elements to form compounds called oxides, such as carbon dioxide.

Most of the oxygen in the air exists as a diatomic molecule, or two atoms that share electrons. However, oxygen can also exist in another form known as ozone. Ozone has three oxygen atoms joined together. In the upper atmosphere, ozone protects Earth from receiving too much of the sun's radiation.

Sulfur is an important element in living things. Bacteria that live near hydrothermal vents on the ocean floor use sulfur compounds to make food.

GROUP 17

Halogens

F Fluorine
Cl Chlorine
Br Bromine
I Iodine
At Astatine

All of the halogens are nonmetals that have seven valence electrons. Because they only need one electron to become stable, they are very reactive. The atoms of halogens combine readily with other atoms, especially those of metals.

Fluorine is the most reactive of all the elements. As a result, fluorine atoms do not exist uncombined in nature. Fluorine is added to water supplies and toothpastes (as fluoride) to prevent tooth decay. Some compounds of fluorine were once used in cooling systems. Those compounds are no longer used because they caused damage to Earth's ozone layer.

Like fluorine, chlorine is never found uncombined in nature. Chlorine is a poisonous green gas with a strong odor. In its liquid and solid forms, chlorine is used to produce paper products, plastics, medicines, and paints.

Small amounts of iodine are needed by the human body. Iodine is part of a hormone that controls the body's metabolism. A lack of iodine can cause illness.

The Root of the Meaning

The word ***halogen*** comes from the Greek word *hals,* meaning "salt," and *genēs,* meaning "former" or "born." Halogens combine with other elements to form salts.

Chlorine is used to treat drinking water and swimming-pool water.

Everyday Science

Table Salt

Every time you add salt to food, you are using a chemical compound. Table salt, also known as sodium chloride (NaCl), is a compound of the elements sodium (Na) and chlorine (Cl). Sodium is an alkali metal and chlorine is a halogen.

GROUP 18

Noble Gases

He Ne Ar Kr Xe Rn

Unlike other elements, all of the noble gases have eight valence electrons. That's a complete set. Because they are already stable, they are nonreactive. They are all colorless, odorless, and tasteless gases.

Helium was the first noble gas to be discovered. It was discovered in the sun before it was identified on Earth. For this reason, it was named after the Greek god of the sun, Helios. Argon is the most abundant of the noble gases. It makes up a little less than one percent of the atmosphere.

Historical Perspective

Noble Gas Compounds

Until 1962, scientists believed that the noble gases could not form compounds. In fact, these gases were called inert, meaning "not active." Then a compound of xenon was synthesized. Once thought to be impossible, compounds of radon and krypton were produced. These breakthroughs provided a better understanding of both noble gases and the nature of atoms and compounds.

◀ Helium is less dense than any other gas except hydrogen. That is why balloons and blimps filled with helium can float in air

▼ The flammability of hydrogen makes it useful as a fuel to lift the space shuttle from Earth's surface.

◀ Neon is a noble gas that can be made to glow in signs.

Hydrogen

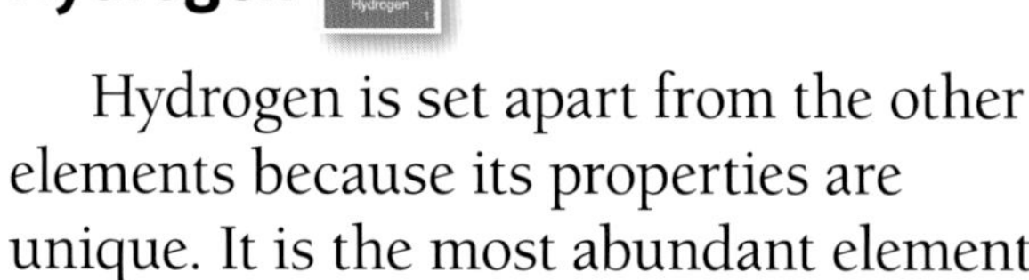

Hydrogen is set apart from the other elements because its properties are unique. It is the most abundant element i the universe. Hydrogen is a nonmetal tha exists as a gas under standard conditions. It is colorless, odorless, and tasteless, but it is very flammable. It is the lightest of all the elements. Hydrogen can form compounds with many other elements. One of the most common compounds of hydrogen is water.

Summing Up

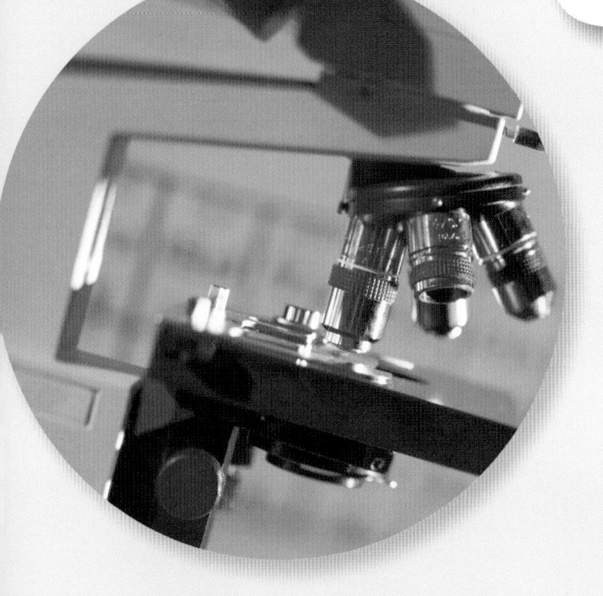

- After contributions by many scientists, the modern periodic table was developed. It arranges the elements in order of increasing atomic number.

When the elements are arranged in this way, their properties repeat at regular intervals. This is known as the periodic law.

Rows of the periodic table are called periods and vertical columns are called groups.

Elements within a group share similar properties because they have the same number of valence electrons.

In addition to groups, elements can be described as metals, nonmetals, and metalloids.

Putting it All Together

Choose one of the following research activities. Answer the question independently, in pairs, or in a small group. Share your responses with the class. Listen to other groups present their answers.

1. Write a brief newspaper article that could have followed Mendeleev's discovery of the periodic law. Be sure to include all the necessary information: who, what, where, when, why, and how.

2. Choose one element from the periodic table. Research its discovery, properties, and uses. Prepare a summary that includes such information as who discovered the element, how it was named, and how it is used. Include any other information you find interesting.

3. Select a group from the periodic table. Find pictures in magazines, books, or newspapers that relate to the elements in the group. Make a poster that describes the group and presents the pictures you find.

Parts Make a Whole

Over thousands of years, scientists developed the modern theory of the atom. Through research and experimentation, they learned to describe atoms by their subatomic particles and to develop models that could be used to understand atoms. The elements were then organized into the periodic table. This table has become a tool that scientists use to understand the properties of elements and make predictions about how elements will behave when combined.

Can you now answer the question posed at the beginning of this book: What does matter have in common with the Great Wall of China? Matter is made up of atoms, about which scientists know a great deal today. Both the Great Wall of China and atoms consist of smaller parts. You can see the parts—the stones and bricks—of the Great Wall, so you know they are there. You can observe them and investigate their properties. The smaller parts of atoms, however, are too small to see or touch. But you can learn about these parts through models that attempt to explain their properties.

How to Write a Biography

A biography is the story of someone's life. A biography can be presented in just a few sentences or it can fill an entire book. A biography should describe and analyze the events in a person's life. It should try to find meaning in actions and make arguments about the significance of the person's accomplishments.

Many biographies are written in chronological order, which means that they follow a time line from earliest events to latest events. Some biographies focus on a specific period of time or a specific event.

"I have no special talent. I am only passionately curious." –A. Einstein

Biographers, or people who write biographies, use two types of sources. Primary sources include letters the person may have written or a journal the person may have kept. Secondary sources include other biographies, reference books. or newspaper accounts that provide information about the person.

Try your hand at writing a short biography. First choose a person you find interesting. Conduct research to learn basic facts about the person's life. Keep track of the sources you use. You might begin with a general resource, such as an encyclopedia, and then discover more specific resources, such as the notes of a great scientist.

Ask the following questions:

- What qualities describe the person?
- What events shaped this person's life?
- What obstacles did the person face? Were they overcome?
- How were other people affected by the person?

Once you know about the person, decide how you will organize the biography. Will you describe the life of the person from start to finish? Will you focus on a single event or series of events? Will you start with one achievement and then work backward?

"A person who never made a mistake never tried anything new." – A. Einstein

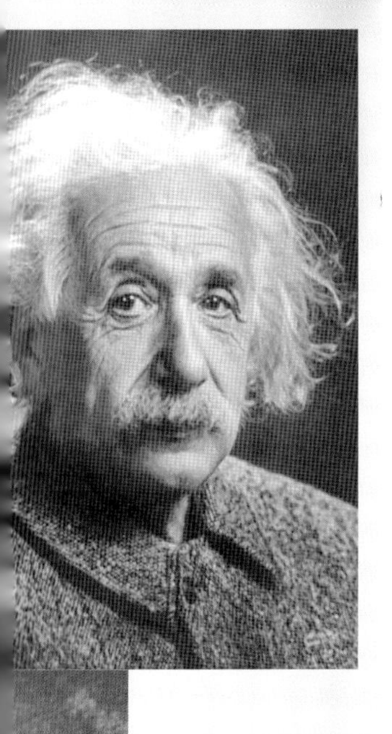

Albert Einstein was born in Ulm, Germany, on March 14, 1879. Soon after his family moved to Switzerland. He was a quiet and shy boy, but he was very smart and had a good sense of humor. He loved to solve puzzles and build things.

Einstein did not like school very much but he was very curious about the world. He liked to ask questions and then find the answers. He was good at math and science, so he trained to be a teacher. In 1901, he graduated from school. He could not find work as a teacher. So, instead, he took a job at the Swiss Patent Office. This is where people patent their ideas and inventions.

While he was at the patent office, and in his spare time, he worked very hard on his own ideas. He had a lot of questions about light and physics. He set out to answer those questions. Soon he began to publish his ideas. He became a college professor in 1901. He married Mileva Maric in 1903 and they had a daughter and two sons. The marriage ended in 1919. Einstein soon remarried Elsa Löwenthal. During this time he wrote many books and traveled around the world teaching people about physics.

In 1933, he and Elsa moved to America. He became a physics professor at Princeton University. Elsa died three years later in 1936. In the years after World War II, Einstein was a leading figure. Though he was famous all over the world, for his ideas, he remained shy all of his life. He spent a lot of time alone, thinking and listening to music. Einstein always appeared to have a clear view of a problem. He was also determined to solve each problem. He had a special way of working. He visualized, or pictured in his mind, the stages on the way to the solution. This helped him reach his goals.

His ideas and scientific discoveries changed what we know about physics. His well-known books include: *Special Theory of Relativity* (1905), *Relativity* (English translations, 1920 and 1950), *General Theory of Relativity* (1916), and *The Evolution of Physics* (1938). He also wrote books about philosophy and peace. Einstein showed the world what hard work and a strong brain could do. He died on April 18, 1955, in Princeton, New Jersey.

Glossary

atom (A-tum) *noun* a building block of matter, made up of protons and neutrons in the nucleus and electrons around the nucleus (page 6)

atomic mass (uh-TAH-mik MAS) *noun* a weighted average of the masses of all the naturally occurring isotopes of an element (page 14)

atomic number (uh-TAH-mik NUM-ber) *noun* the number of protons or electrons in the nucleus of an atom (page 14)

electron (ih-LEK-trahn) *noun* a negatively charged subatomic particle that travels around the nucleus (page 10)

electron cloud (ih-LEK-trahn KLOWD) *noun* a region in which electrons are likely to be found around the nucleus of an atom (page 12)

element (EH-leh-ment) *noun* a pure substance made up of atoms with the same atomic number (page 14)

group (GROOP) *noun* a vertical column, or family, of the periodic table (page 27)

isotopes (I-suh-topes) *noun* atoms of the same element that have the same number of protons (atomic number) but a different number of neutrons (page 16)

mass number (MAS NUM-ber) *noun* the sum of the protons and neutrons in an atom (page 16)

matter (MA-ter) *noun* anything that has mass and takes up space (page 6)

metal (MEH-tul) *noun* an element that is shiny, conducts heat and electricity well, and is malleable and ductile (page 30)

metalloid (MEH-tul-oid) *noun* an element that is neither a metal nor a nonmetal, but shows some of the properties of each (page 31)

neutron (NOO-trahn) *noun* a subatomic particle that is electrically neutral and is located within the nucleus (page 11)

nonmetal (nahn-MEH-tul) *noun* an element that is not shiny, does not conduct heat and electricity well, and is not malleable or ductile (page 31)

nucleus (NOO-klee-us) *noun* the center of an atom, which contains protons and neutrons (page 11)

period (PEER-ee-ud) *noun* a horizontal row of the periodic table (page 27)

periodic law (peer-ee-AH-dik LAW) *noun* an understanding that the chemical and physical properties of elements repeat in a pattern when the elements are arranged in order of increasing atomic number (page 27)

proton (PROH-tahn) *noun* a subatomic particle with a positive electric charge that is located within the nucleus (page 11)

valence electron (VA-lents ih-LEK-trahn) *noun* an electron in the outmost energy level of an atom (page 33)

Index